Our Exciting Earth!

SWAMPS

By Tanner Billings

Gareth Stevens
PUBLISHING

Please visit our website, www.garethstevens.com.
For a free color catalog of all our high-quality books, call toll free 1-800-542-2595 or fax 1-877-542-2596.

Library of Congress Cataloging-in-Publication Data
Names: Billings, Tanner, author.
Title: Swamps / Tanner Billings.
Description: New York : Gareth Stevens Publishing, 2023. | Series: Our exciting Earth! | Includes index.
Identifiers: LCCN 2021043467 | ISBN 9781538275764 (set) | ISBN 9781538275771 (library binding) | ISBN 9781538275757 (paperback) | ISBN 9781538275788 (ebook)
Subjects: LCSH: Swamps–Juvenile literature. | Swamp ecology–Juvenile literature.
Classification: LCC GB621 .B55 2023 | DDC 577.68–dc23/eng/20211008
LC record available at https://lccn.loc.gov/2021043467

First Edition

Published in 2023 by
Gareth Stevens Publishing
29 East 21st Street
New York, NY 10010

Editor: Kate Mikoley
Designer: Tanya Dellaccio

Photo credits: Cover Ezume Images/Shutterstock.com; p. 5 Anda Mikelsone/Shutterstock.com; p. 7 Vad-Len/Shutterstock.com; p. 9 K64End/Shuttertock.com; p. 11 red rose/Shutterstock.com; p. 13 Georgios Antonatos/Shutterstock.com; p. 15 Sasha Craig/Shutterstock.com; p. 17 alybaba/Shutterstock.com; pp. 19,21 jaimie tuchman/Shutterstock.com; p. 23 lazyllama/Shutterstock.com.

Printed in the United States of America

CPSIA compliance information: Batch #CSGS23: For further information contact Gareth Stevens, New York, New York at 1-800-542-2595.

Contents

Some land is always wet.
It's a swamp!

Water covers a
swamp at times.

The water wets the soil.
It moves slowly.

Some have fresh water.
Some are salty.

Swamps are often
by a river.
Some are by the sea.

A big swamp is in Louisiana.

Swamps have many trees.
Gum trees are one kind.

Many animals live here. Some are fish, frogs, and birds.

Bugs live here too.

Swamps are all over. There may be one near you!

Words to Know

bird

soil

Index